Night Sky

Written by
CAROLE STOTT

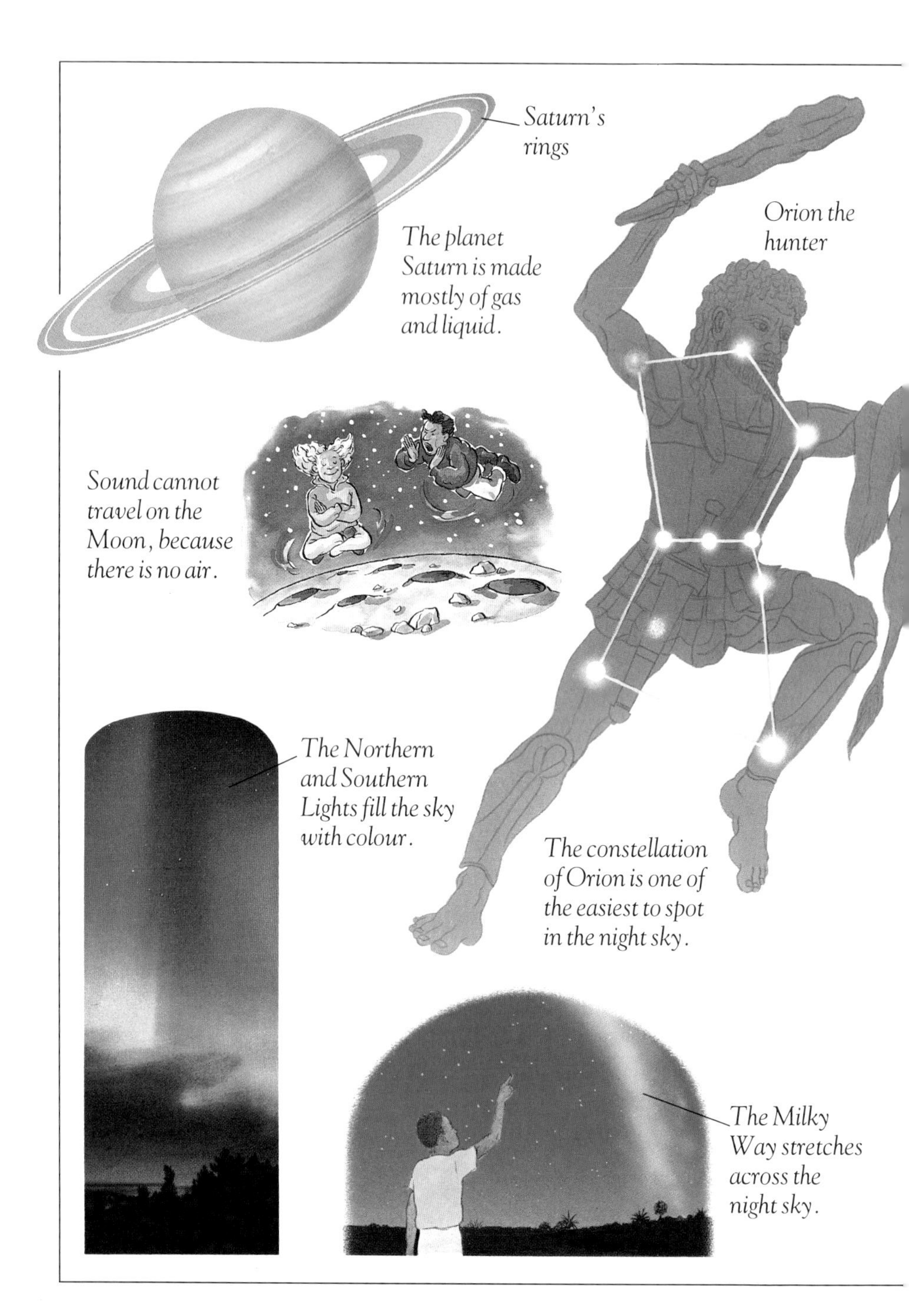
Saturn's rings
The planet Saturn is made mostly of gas and liquid.
Orion the hunter
Sound cannot travel on the Moon, because there is no air.
The Northern and Southern Lights fill the sky with colour.
The constellation of Orion is one of the easiest to spot in the night sky.
The Milky Way stretches across the night sky.

Night Sky

Written by
CAROLE STOTT

DORLING KINDERSLEY
London • New York • Stuttgart

A DORLING KINDERSLEY BOOK

Editor Bernadette Crowley
Art editor Vicky Wharton
Senior editor Susan McKeever
Production Catherine Semark

This Eyewitness ® Explorers book
first published in Great Britain in 1993 by
Dorling Kindersley Limited
9 Henrietta Street
Covent Garden
London WC2E 8PS

A CIP catalogue for this book is available from the British Library.

ISBN 07513 6100 3

Colour reproduction by Colourscan, Singapore
Printed in Spain by Artes Gráficas Toledo, S.A.
D.L.TO: 854-1996

The author would like to dedicate
this book to Ellen

Visit us on the World Wide Web at
http:/ /www.dk.com

Contents

Looking at the night sky

Wherever you live in the world, you can look up at the night sky and see the Moon and the twinkling stars. But how many stars you see depends on the weather, and where you are looking from.

Starry night

It is best to look at the night sky away from street and house lights, which block out starlight. You will see the most stars from the countryside, where the sky is darkest. But if you live in a city, you can still see plenty of stars.

Night sky notebook

Make notes and draw what you see in the sky. Keep a record of the date, the time, and your location.

Don't go out alone at night – always ask an adult to go with you.

It can be very cold at night, so wrap up well. Remember to bring something to sit on, as well as something to eat and drink.

Be careful when using scissors – they are sharp!

Night vision

Get your eyes used to the dark before you look at the night sky. Switch off the house lights, but stay still until you can see where you are going or you might bump into something!

Seeing stars

Once outside, you will need a torch to see your notebook. Cover your torch with red cellophane. A red light will not spoil your eyesight once it is used to the dark.

1 Using scissors, cut a piece of red cellophane large enough to cover the end and a little of the side of the torch.

You will need sticky tape, scissors, and red cellophane.

2 Use the sticky tape to stick the cellophane to the torch. A rubber band will also do the job.

You can gaze at the stars through your bedroom window. Make sure you turn the lights out first or you won't be able to see the stars!

What's in the night sky?

There is more than you might think in the night sky. You may recognize the Moon and stars, but would you know a planet, a galaxy, or a star "nursery" if you saw one? We can see planets, distant galaxies, and much, much more. It is just a matter of knowing what they look like and where to find them.

The Moon does not always look the same. At times we see only part of it.

Some planets have rings around them.

Comets visit the Earth's sky from time to time.

There are eight other planets besides Earth. You can see some of them in the night sky; they look like bright, round stars.

Most of the objects in the night sky are stars.

The Sun is our nearest star. It lights up the daytime sky and blots out everything else. The stars, planets, and galaxies are still in the sky. We just can't see them.

Spaced out

The Moon is the closest object to Earth, so we can see it very clearly. Other objects you see in the night sky are a lot bigger than the Moon, but because they are so much further away they only look like dots of light to us.

Bang!

Everything we know of lives in the Universe. It began about 15,000 million years ago with a huge explosion called the Big Bang. The explosion pushed the young Universe apart. Over millions of years the material formed the galaxies, the stars, the planets, and finally, the human race.

A nebula is a cloud of gas and dust. Stars are being born in this nebula.

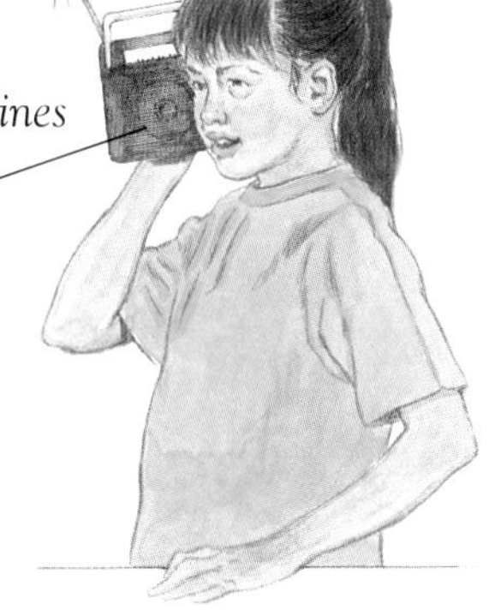

The hiss you hear is radio energy from Earthly things, such as car engines and lightning.

Because nebulae are so far away, they look like hazy patches of light in the Earth's sky.

Sounds of the Universe

Everything in the Universe, including you, produces energy. One type of energy is light, another is heat, and another is radio energy. Scientists use special equipment to hear the radio energy coming from stars and galaxies. When you change stations on a radio, the hiss you hear between stations is radio energy coming from things on Earth.

Millions of stars live together in galaxies. They too look like hazy patches of light.

What is a star?

There are billions of stars in the Universe. Stars are gigantic balls of glowing gas, which live for millions and millions of years. There are two gases inside a star – hydrogen and helium. A star uses its gases to produce heat and light. The Sun is a star, and we feel its heat and light on Earth.

In the beginning
Stars are born in spinning clouds made of gas with a little bit of dust mixed in. A force called gravity pulls the gas and dust together, forming small clouds. The spinning clouds then shrink and form themselves into ball shapes.

Star turns
These spinning balls of gas and dust are stars at the start of their lives. If a ball spins slowly it will produce one star; if it spins fast, twin stars are born. Medium spin will produce a star with planets: this is what happened when the Sun was born.

Stars are beginning to form.

Colours of light

Light is made of different colours. When light shines through water, it splits into these colours. To split the Sun's light, you'll need water in a straight glass, a piece of card, and some white paper. Cut a long, narrow hole in the card. Place the glass on the paper by a sunny window. Sit the card between the window and the glass. You can now see the colours on the paper.

How many colours can you see?

Pull and push

Everything in the Universe has its own gravity. Gravity is a pulling force that attracts objects towards one another. Inside a star, gravity battles with a pushing force called pressure. As gravity pulls the gas in, pressure pushes it out, so the gas in a star keeps its ball shape.

Young shiners

The young stars start to turn their gases into heat and light and slowly begin to move apart.

Sisterly stars

The Pleiades (*ply-ay-dees*) is a group of around 300 very young stars. It is also known as the Seven Sisters because seven of its stars are visible to the naked eye.

Sparklers

A group of stars called the Jewel Box has many colourful stars. You would need a powerful telescope to get this view (left). It looks hazy to the naked eye.

Life of a star

Stars live for millions of years. As a star grows older, it changes. The Universe contains young, middle-aged, and old stars. The biggest stars are 100,000 times bigger than the Sun, and the smallest stars are much smaller than Earth.

Star colours

All stars are incredibly hot, but some are hotter than others. Blue-white stars are the hottest of all. Yellow-orange ones, like the Sun, are cooler. Red stars are coolest of all. But even red stars are 20 times hotter than a kitchen oven.

This star is middle-aged, like our Sun. But its yellow-white colour tells us it is hotter and brighter.

View through binoculars shows that "single" star is really a double star.

Seeing double

About half the stars in the Universe are double stars. These are two stars that live together, or two stars that look close together because we are looking at them from such a long way away. Use binoculars to look for double stars.

Red giant

Different stars change in different ways. How a star changes depends on how much gas it is made of. When stars like the Sun have used up all their hydrogen, their surface will cool and turn red. They will swell up and become red giants. But don't worry; this won't happen to the Sun for another 5,000 million years!

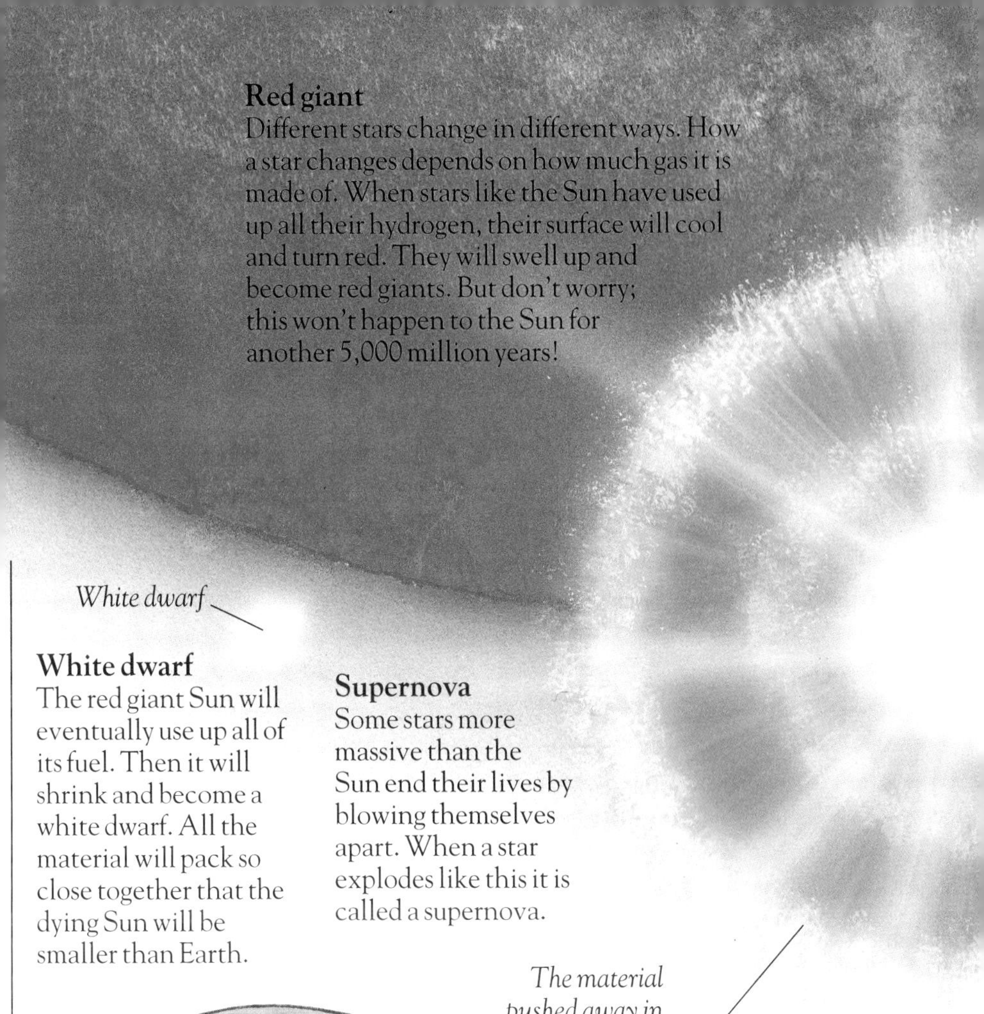

White dwarf

White dwarf

The red giant Sun will eventually use up all of its fuel. Then it will shrink and become a white dwarf. All the material will pack so close together that the dying Sun will be smaller than Earth.

Supernova

Some stars more massive than the Sun end their lives by blowing themselves apart. When a star explodes like this it is called a supernova.

The material pushed away in the explosion will produce new stars.

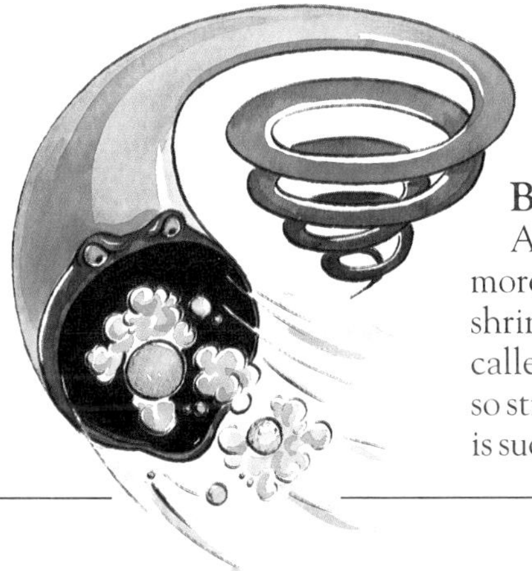

Beware of the black hole!

As some stars die, their material gets more and more squashed together. The star shrinks until it is a point in space. This is called a black hole. A black hole's gravity is so strong that anything that gets too close is sucked into it – and can never escape.

Dots of light

As we look out into space, we see lots of twinkling dots of light. These dots are huge stars, but because they are so far away they just look like bright specks in a dark sky. It seems as if the Earth is inside a gigantic sphere covered in stars.

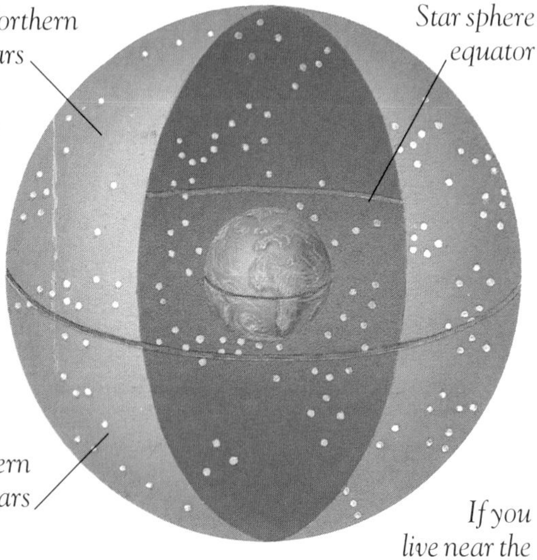

Northern stars

Star sphere equator

The dividing line

The Equator, an imaginary line around the middle of Earth, divides our world in two. The top half is the Northern Hemisphere and the bottom half is the Southern Hemisphere. The sphere of stars also has an equator dividing it into northern and southern halves.

Southern stars

You can see mainly northern stars from the Northern Hemisphere and mainly southern stars from the Southern Hemisphere.

If you live near the Equator, you can see stars from both halves of the star sphere.

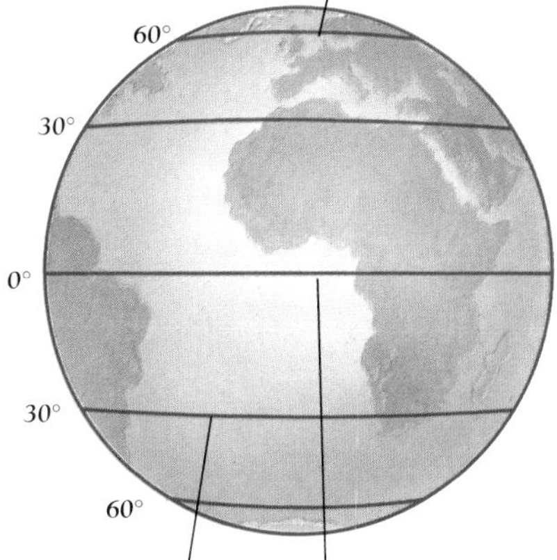

This latitude line is 60° north (of the Equator).

60°

30°

0°

30°

60°

This latitude line is 30° south (of the Equator).

The Equator is 0° latitude.

Which line are you on?

Imaginary lines called latitude lines help us describe where a place is on the Earth's surface. They are counted north or south of the Equator and are measured in degrees (°). Knowing which latitude line you live on will help you know which stars are in the sky above your home.

Finding directions

To study the stars, you'll need to know in which direction to look. Make a compass with a stick and some stones.

1 Push the stick into the ground. As the Sun shines on the stick, it will make a shadow.

The shortest shadow points north if you live in the Northern Hemisphere, and south if you live in the Southern Hemisphere.

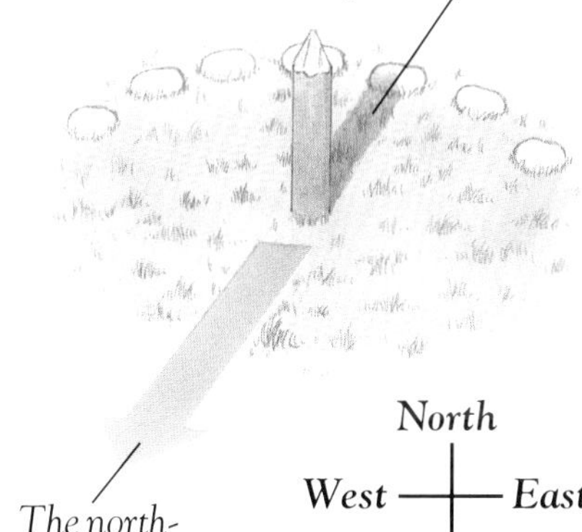

The north-south line.

2 Place your stones at the end of the shadow at different times during the day – the first one in the morning and the last one in the afternoon. The stones will make a curve. The north-south line is where the Sun made the shortest shadow. The east-west line cuts at right angles to it.

North

West — **East**

South

The four main compass points

Twinkling stars

Starlight shines steadily. But stars twinkle in our sky. This is because their light wobbles as it passes through the Earth's atmosphere.

Light wobbles in the Earth's atmosphere (the layer of gas that surrounds the Earth).

Travelling light

Light travels faster than anything else, but stars are so far away from us that their light takes years to reach Earth. Astronomers (scientists who study the stars and planets) can see very distant galaxies whose light started travelling towards us when dinosaurs lived on Earth, millions of years ago.

Patterns in the sky

When humans first looked at the sky, they saw hundreds of twinkling dots of light. They joined up the dots to make pictures of people, animals, and other things. The pictures helped them, and now us, to remember the stars.

Orion

A group of stars and its picture is called a constellation. This is the constellation of Orion, the hunter. It is one of the easiest constellations to see in the sky.

Star light, star not so bright

Some stars look brighter than others. Over 2,000 years ago, Greek astronomer Hipparchus numbered the stars 1, 2, 3, 4, 5, or 6. The brightest stars were numbered 1 and the dimmest were numbered 6. Astronomers still use his numbering system today.

A bright heart

The Lion is another constellation. The Lion's heart is marked by the bright star Regulus. A lion cub cuddles close to the Lion, but its stars are not so bright and it is difficult to see.

Make a model of Orion

The stars in a constellation are different distances away from us. If we could see Orion from somewhere else in space, it would have a different pattern. Make a model of Orion using a shoe box, seven straws, modelling clay, and the picture of Orion on page 18.

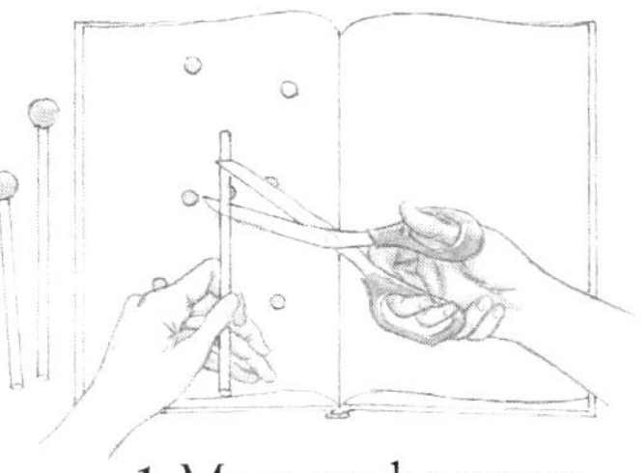

1 Measure the straws from the bottom of the page to the two stars in Orion's shoulders; the three stars in his belt, and the two stars on his thigh and ankle. Cut one straw for each star. Mould one small ball of modelling clay for each star and stick a ball on to the end of each straw.

2 Cut a window in one end of the shoe box. Turn the box on its side. Arrange the stars inside the box so that when you look through the window the stars show the pattern of Orion. (Stand the straws in small pieces of modelling clay.)

3 You can see Orion's pattern through the front window, but from the side, you can see how your stars make a completely different pattern. This is how Orion might look from somewhere else in the Milky Way.

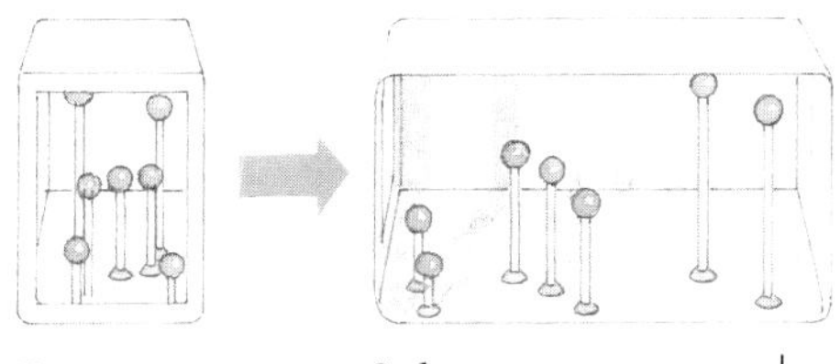

Front view *Side view*

Northern stars

If you live in the Northern Hemisphere, look for the constellations on this map. You cannot see all the constellations in one night. The stars towards the edge of the map can be seen only at certain times of the year, but those in the centre are always visible.

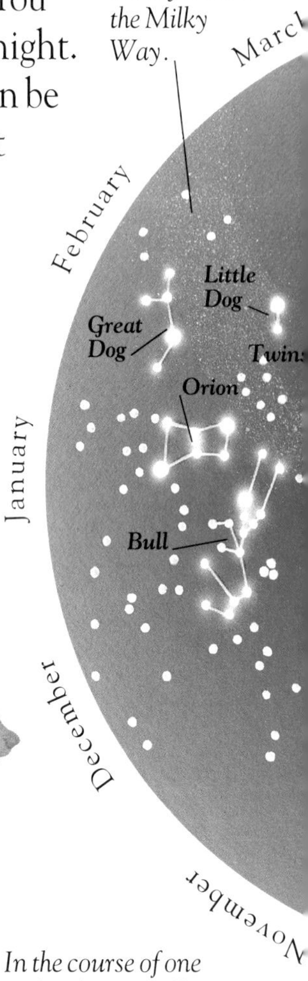

The milky river of starlight that stretches across the sky is called the Milky Way.

Northern sky map

To use this map, turn it so that the current month is at the bottom. The stars in the middle and on the lower half of the map are the stars you can see in that month.

The most visible part of the Great Bear is the seven stars in the shape of a pan, known as the Plough.

The bears up there

There are two bears in the sky – the Great Bear and the Little Bear. These two constellations can be seen all year round. Cepheus (*see-fee-us*), Cassiopeia (*cass-ee-o-pee-uh*), and the Dragon can also be seen throughout the year.

In the course of one night, the constellations change their position in the sky. One star – Polaris – stays still, while the other stars circle around it.

At first, you may find it difficult to find these constellations, but it will become easier with practice.

Heavenly queen

In an ancient Greek myth, Cassiopeia, queen of Ethiopia, angered the sea god Poseidon, so he banished her to the sky forever. As Cassiopeia moves around Polaris, she spends some of the time upside-down!

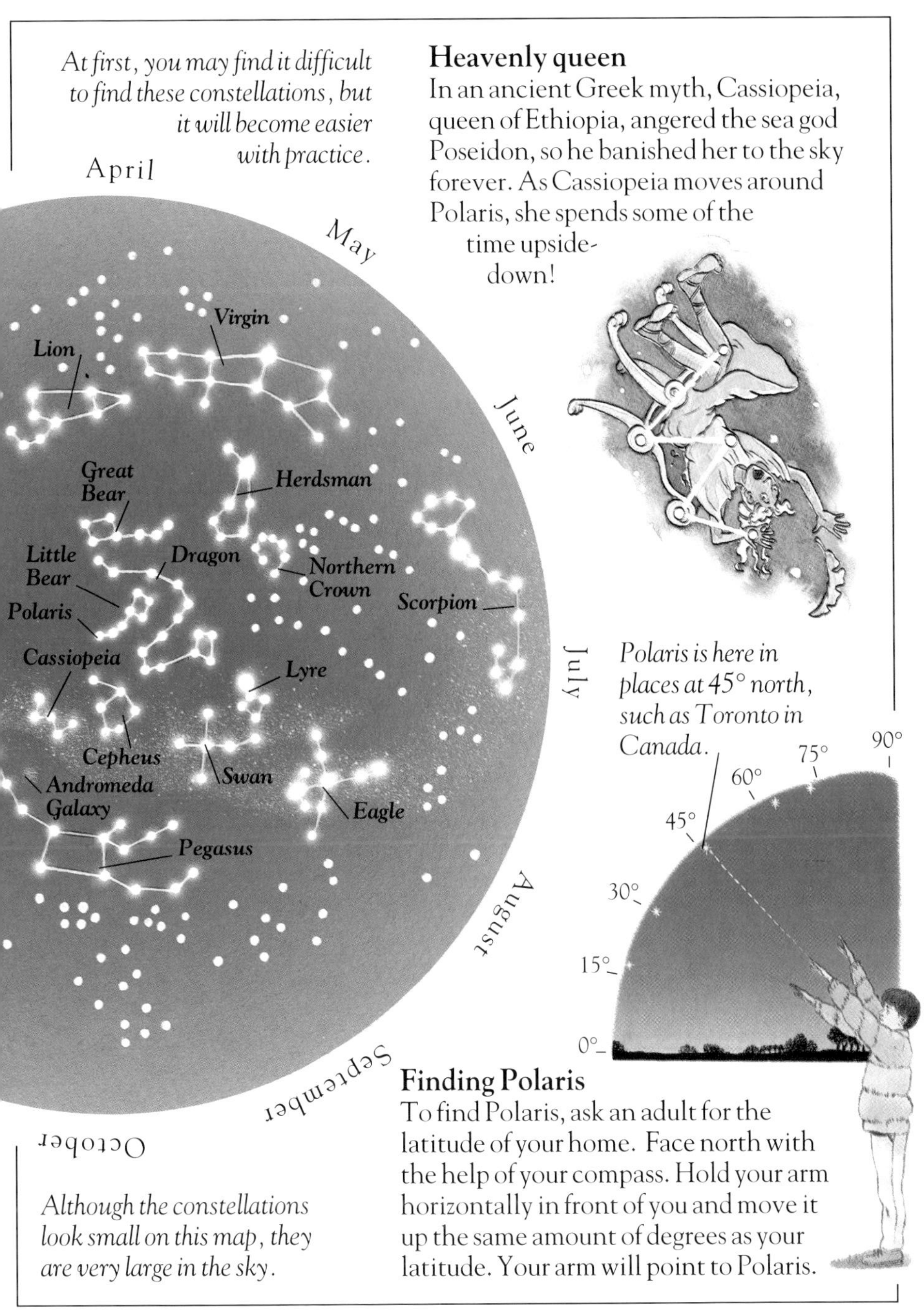

Polaris is here in places at 45° north, such as Toronto in Canada.

Although the constellations look small on this map, they are very large in the sky.

Finding Polaris

To find Polaris, ask an adult for the latitude of your home. Face north with the help of your compass. Hold your arm horizontally in front of you and move it up the same amount of degrees as your latitude. Your arm will point to Polaris.

Southern stars

Can you find which constellations are also on the northern sky map?

The southern sky is packed with all sizes of constellations. The smallest of all is the Cross. It is visible all year round, just like the other constellations in the centre of the map. The constellations at the edge of the map come and go with the seasons.

How to use this map

Turn the map so that the current month is at the bottom. The constellations in the middle and on the lower half of the map are the ones you will be able to see in that month.

The constellations around the edge of the map can be seen in both hemispheres.

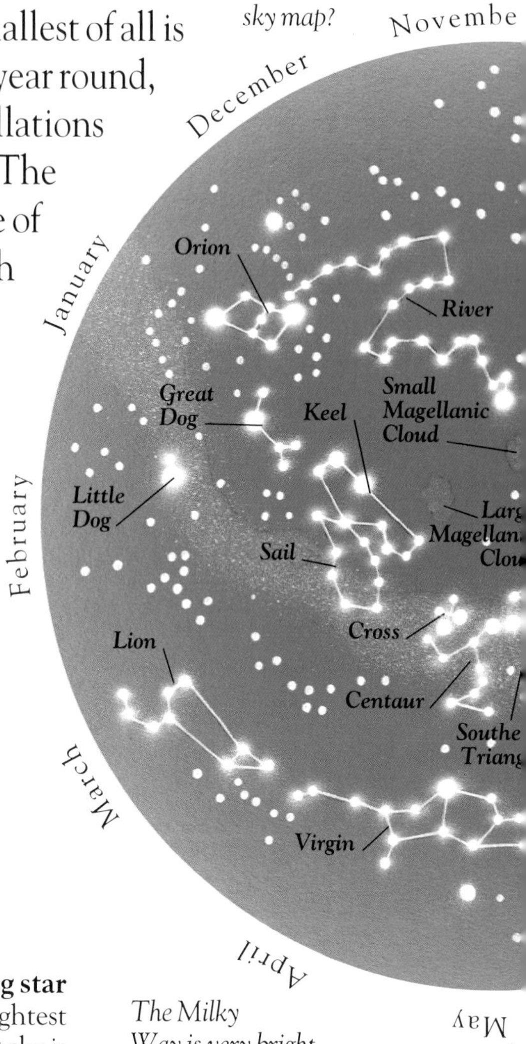

Dog star

The brightest star in the night sky is Sirius. It is also called the Dog Star because it belongs to the star constellation called the Great Dog.

The Milky Way is very bright in the southern sky because it contains more stars than the part that crosses the northern sky.

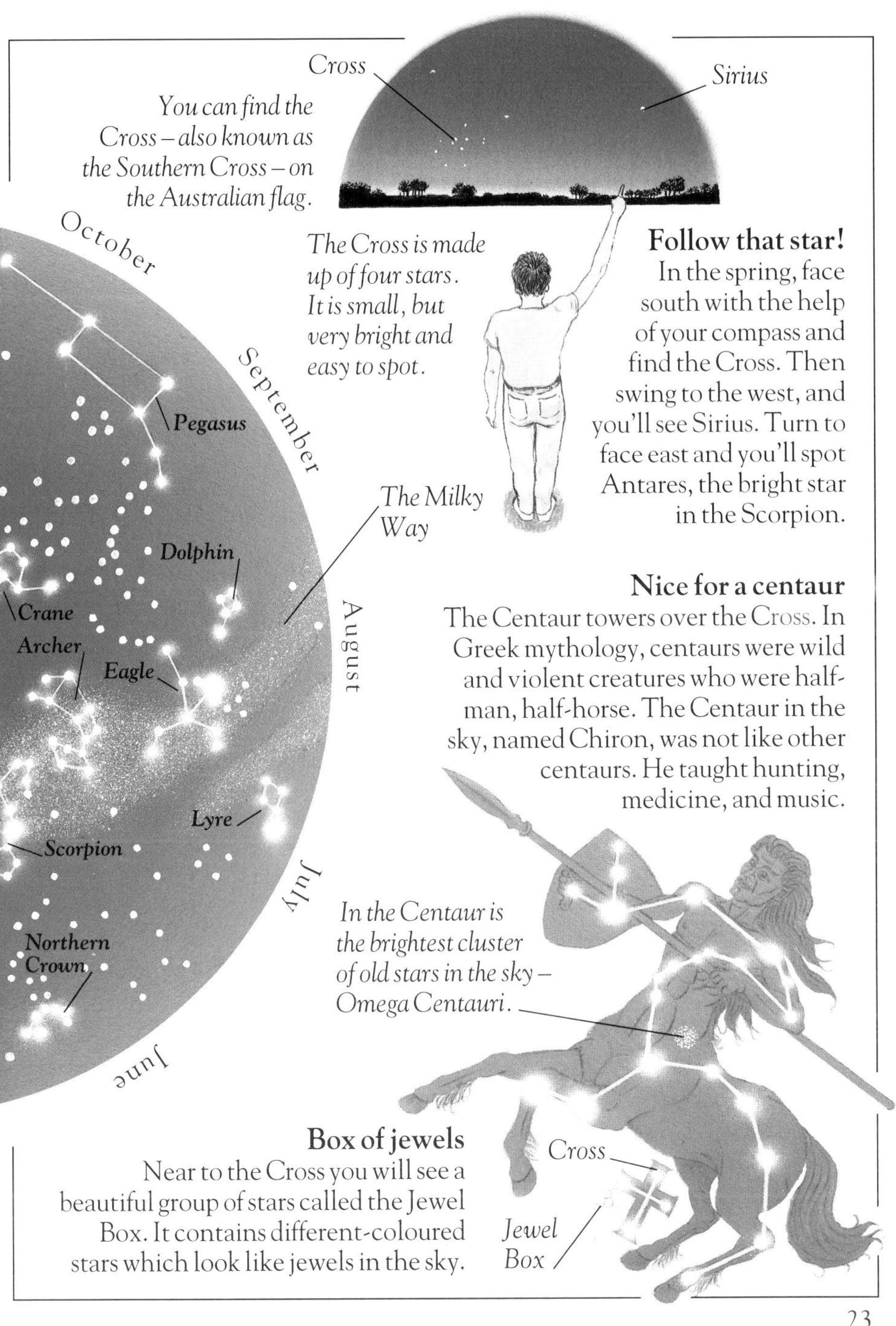

Follow that star!

In the spring, face south with the help of your compass and find the Cross. Then swing to the west, and you'll see Sirius. Turn to face east and you'll spot Antares, the bright star in the Scorpion.

Nice for a centaur

The Centaur towers over the Cross. In Greek mythology, centaurs were wild and violent creatures who were half-man, half-horse. The Centaur in the sky, named Chiron, was not like other centaurs. He taught hunting, medicine, and music.

Box of jewels

Near to the Cross you will see a beautiful group of stars called the Jewel Box. It contains different-coloured stars which look like jewels in the sky.

The zodiac

The zodiac is a circle of twelve special constellations. The first people on Earth saw that as the Sun, the Moon, and the planets moved through the sky, their paths always crossed the same starry background. They divided this background into twelve constellations, which we now call the zodiac.

The zodiac appears in both hemispheres.

The Sun moves from one zodiac constellation to another every four weeks or so.

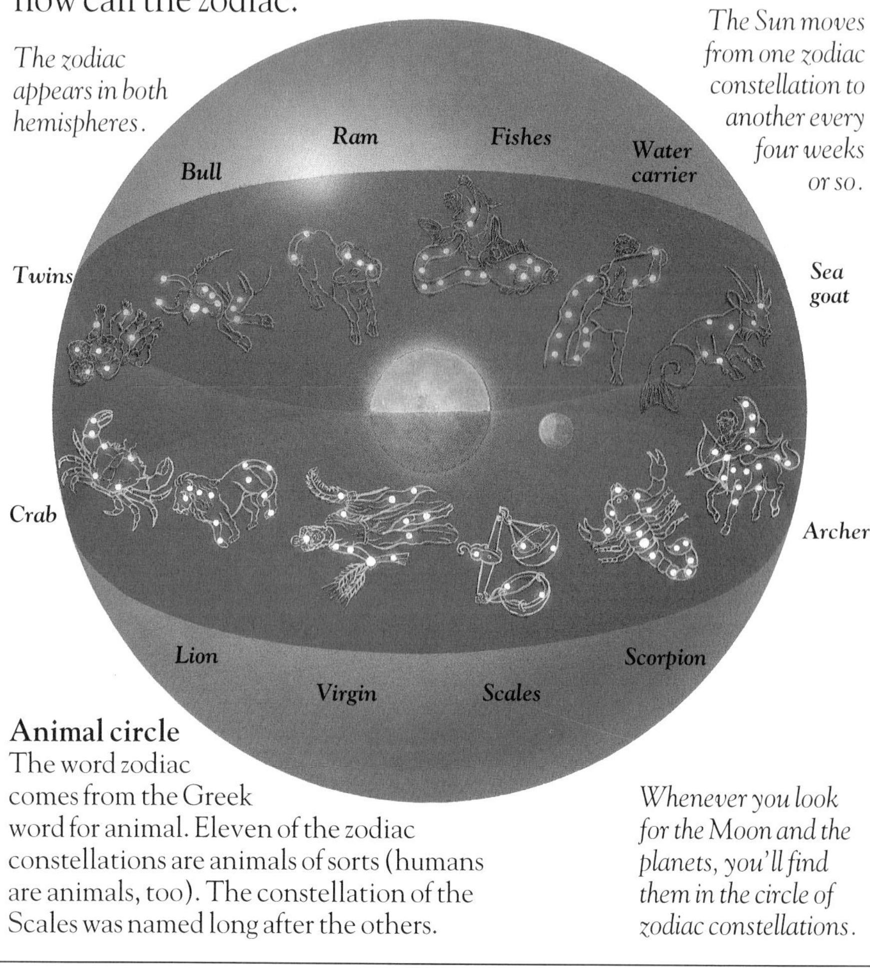

Animal circle

The word zodiac comes from the Greek word for animal. Eleven of the zodiac constellations are animals of sorts (humans are animals, too). The constellation of the Scales was named long after the others.

Whenever you look for the Moon and the planets, you'll find them in the circle of zodiac constellations.

The Scorpion

A scorpion is a small and dangerous creature. In Greek mythology, this scorpion stung Orion to death.

A curving line of stars marks the Scorpion's tail, where it keeps its dangerous sting.

Sting

Antares

Tail

Weather forecast

Astrologers believe that the positions of the Sun, Moon, and planets in the zodiac can tell us about the future. Years ago, farmers would ask an astrologer for a weather forecast!

The Bull

Some star patterns can easily be turned into a picture. In the constellation of the Bull, a V-shape of stars forms his face, and then stretches out to make his horns.

Pleiades

Shine a light behind the card and see the stars shine.

Star patterns

Use the zodiac pictures to learn the star patterns. Choose a constellation and draw its stars on a piece of card. Using a drawing pin, make a hole in each star. Now draw the zodiac picture around the stars.

Once you have drawn the zodiac picture, see if you can make up your own picture.

Fuzzy objects

In the night sky, stars look like twinkling points of light and planets look like small bright discs. But you can also see lots of fuzzy dots and patches of misty light in the sky. These may be galaxies which look misty because they are so far away, or they could be clusters of stars or nebulae which are much closer.

The Orion Nebula is a huge cloud of gas and dust where stars are born. The nebula is lit up by the young stars in it.

Taking a closer look
Some of the fuzzy objects we see are not really fuzzy. Clusters of young stars like the Pleiades look fuzzy to your eyes, but you can see they are a group of stars if you use binoculars. But some objects stay fuzzy even when viewed through very powerful telescopes. These fuzzy objects are nebulae.

The word nebula means "misty". Look at the sword in the constellation of Orion and you'll see a misty patch (see page 18). This is the Orion Nebula.

You can find Hercules between the constellations of the Lyre and the Northern Crown.

Planetary nebulae

Fuzzy objects such as the Ring Nebula are called planetary nebulae, but they have nothing to do with planets. They are shells of gas blown off by old and dying giant stars.

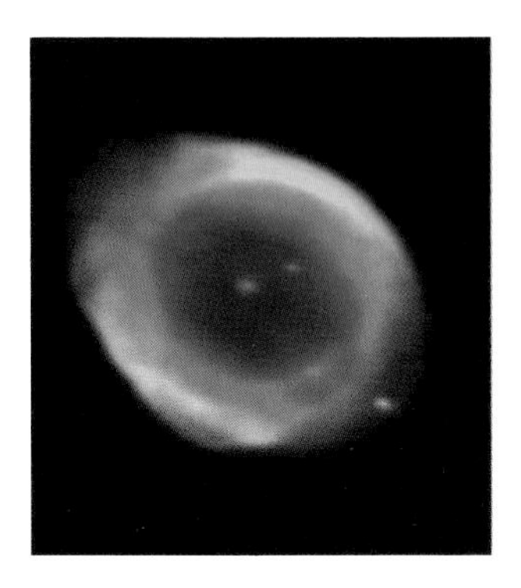

Huge balls of stars

Globular clusters are ball-shaped groups of old stars. They contain hundreds of thousands of stars and look like hazy patches of light in our sky. In the northern sky, the constellation of Hercules has a bright globular cluster.

The globular cluster in Hercules is the brightest in the northern sky.

Can you guess what the three nebulae in this cartoon might be called?

Brilliant bunch

The brightest globular cluster in the sky is in the constellation of the Centaur in the southern sky (see page 23). It will look like a hazy patch of light. But you can see some of its stars with binoculars.

Lookalikes

Over 200 years ago the French astronomer Charles Messier started a list of fuzzy objects in the sky. Some of them have names which describe the way they look.

Galaxies

Stars live together in gigantic groups called galaxies which spin and travel through space. Each galaxy holds many millions of stars and there are millions of galaxies in the Universe. Between galaxies there is only empty space. There are four different galaxy shapes: spiral galaxies, barred spiral galaxies, elliptical galaxies, and irregular galaxies.

Birth of a galaxy
A galaxy starts its life as an enormous spinning cloud of gas and dust. The bigger the cloud, the bigger the galaxy. As the cloud spins, it starts to change shape and stars begin to form.

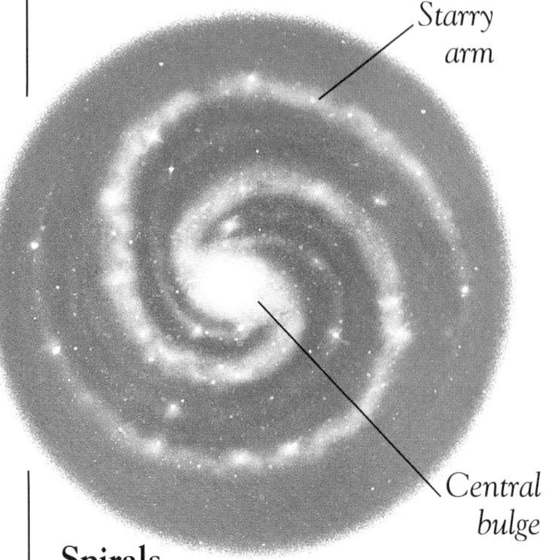

Spirals
Spiral galaxies are in the shape of a flat disc. The stars form arms which spiral out from a big bulge of stars in the centre. All types of stars and nebulae live in spiral galaxies.

Irregulars
Some galaxies have no particular shape at all. These are called irregular galaxies. They are the rarest type of galaxy.

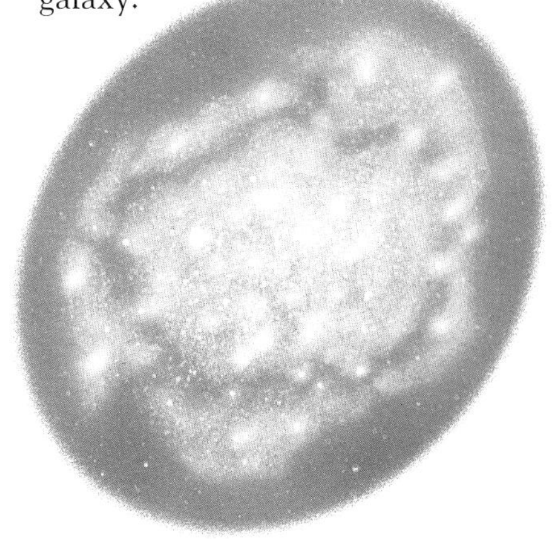

This is the Small Magellanic Cloud in the southern sky.

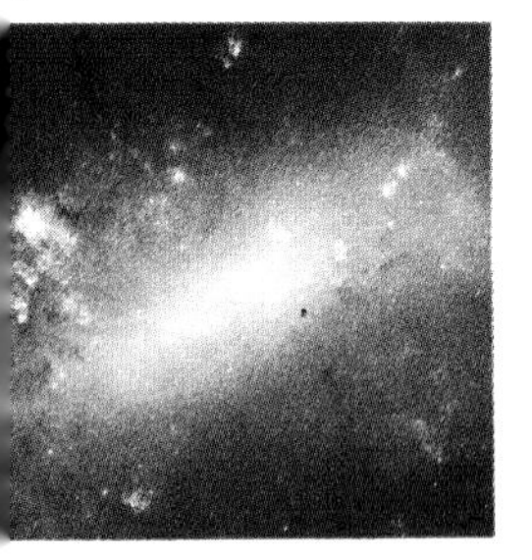

The Andromeda Galaxy in the northern sky is the most distant object you can see without a telescope.

Galaxy gazing

You'll need to look carefully to find a galaxy, as it will be very small in the sky. You can see the galaxies shown above, but get your eyes used to the dark first. The star maps on pages 20-23 will show you where to find them.

Magellan

Early astronomers lived in Europe and Asia and could see only the northern sky. Ferdinand Magellan, a Portuguese explorer, sailed south five hundred years ago and saw two galaxies. They are now called the Small and Large Magellanic Clouds.

Elliptical

Some galaxies are round like balls. Others are more like squashed balls. Some look as though they have been squashed flat. All of these are elliptical galaxies.

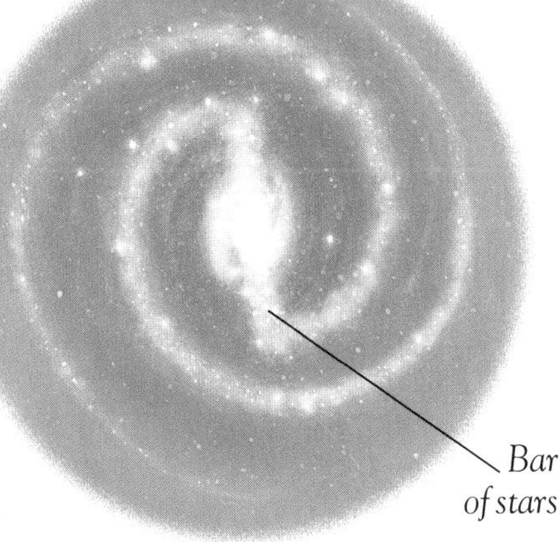

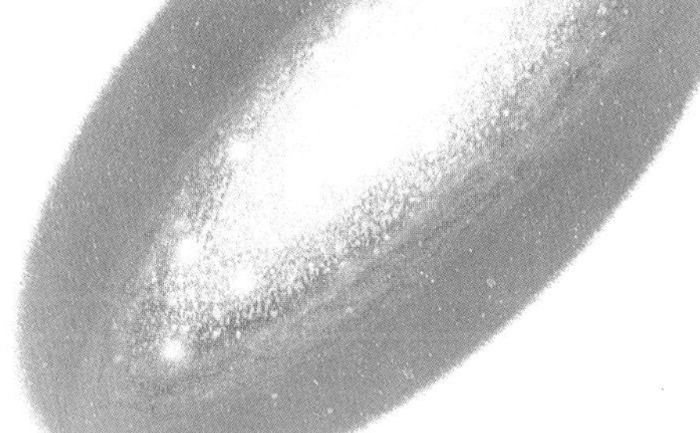

Barred spirals

In this type of spiral galaxy, the stars form a bar in the centre. The arms of stars curl out from both ends of the bar.

The Milky Way

We live in a galaxy called the Milky Way. It is a spiral galaxy with arms swirling out from its centre. There are more stars than you could ever count in our Galaxy – around 500,000 million. All the stars that you see in the night sky are in the Milky Way.

All types of stars live in the Milky Way – young, old, big, and small.

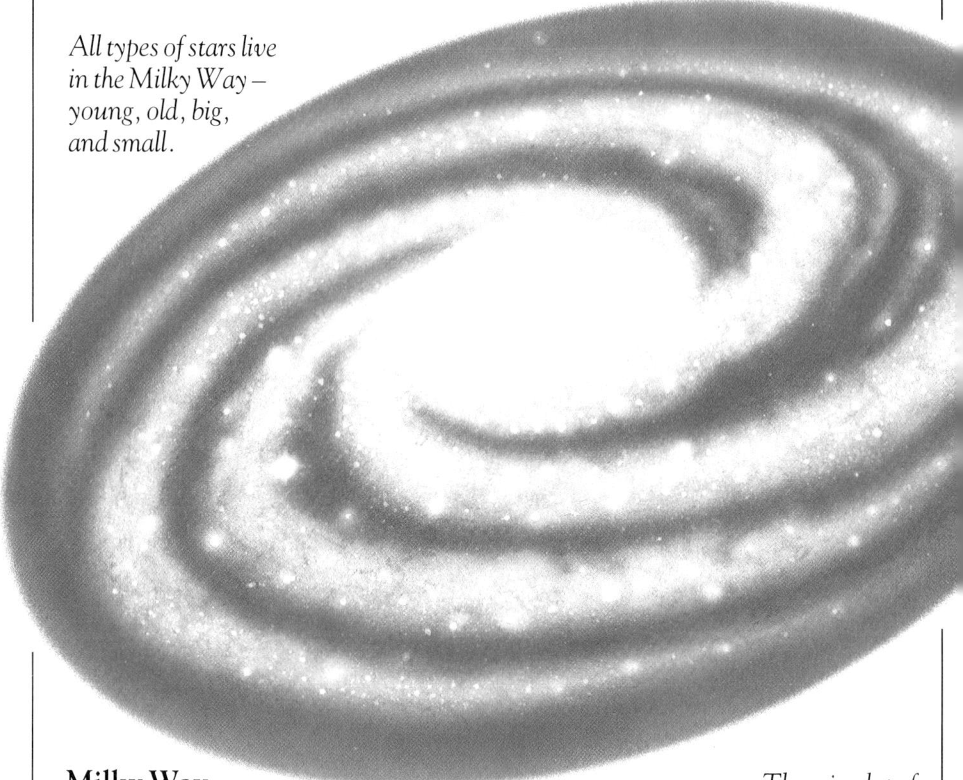

Milky Way
If we could fly above the Milky Way, we would see its spiral shape. But our Galaxy is so large that it would take thousands of years to travel to its edge – even in the fastest rocket.

There is a lot of movement in the Milky Way. The stars travel around the Galaxy's centre, and the whole Galaxy spins in space.

Just one of the stars

Although the Sun is special to us, it is just another star in the Galaxy. The Sun lives in one of the spiral arms, about two-thirds of the way out from the Galaxy's centre.

Spilt milk

The Ancient Greeks thought the Milky Way was milk spilt by the baby Heracles (Hercules) as he drank from the breast of Hera, the queen of the heavens.

From the Earth, you can look either into the centre of the Milky Way, or out to its edge.

Side view

From out in space, the Milky Way looks like a huge pancake. A side view shows the bulge at its centre where most of the stars live. From Earth, you see a band full of stars stretching across the night sky.

Side view of the Milky Way

Milky Way – southern sky

Wherever you live in the world, you can see the path of the Milky Way. It is most spectacular in the sky above the southern half of Earth. Here you can look to the Galaxy's centre.

Milky Way – northern sky

You can see the path of the Milky Way with only your eyes – but binoculars will show you that the milky light is made of thousands and thousands of stars.

The Sun and its family

The star we know best is the Sun. Like all other stars in the sky, the Sun is an enormous ball of hot, glowing gas. It is the only star we know of, out of all the billions and billions of stars, to have its own "family" of planets.

Gigantic clouds of hot gas called prominences leap out from the Sun.

Close for comfort
We are nearly 150 million km (93 million miles) away from the Sun – that's close enough for us to learn a lot about it. If we were much closer we would be roasted alive. The Sun is the only star we can see in detail; astronomers use special equipment to study it. By learning about the Sun, we learn more about the other stars in the Universe.

Far-flung family
The Sun and its family of planets were born from the same cloud of gas and dust. But they are so far apart that it would take the fastest passenger plane about 650 years to visit the most distant family member, Pluto.

Spots often appear on the Sun's face. They are areas of cooler, dark gas.

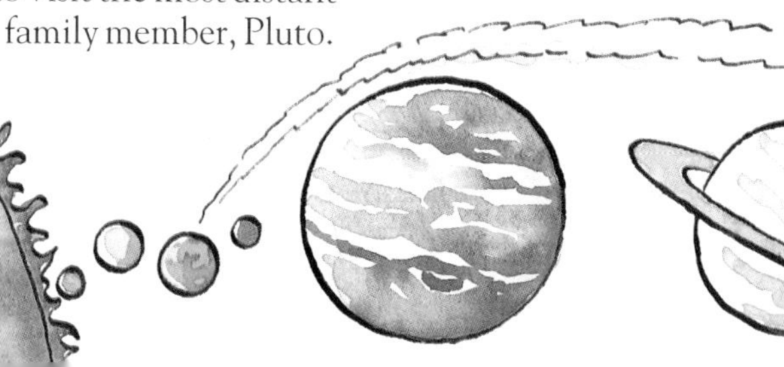

See the Sun

It is dangerous to look directly at the Sun – its bright light will damage your eyes. But there is a safe way to see the Sun. You will need a pair of binoculars, white paper, thin card, and sticky tape.

Take care when you use scissors!

1 Place the binocular eyepieces on the card and draw around them. Cut out the two circles and push the eyepieces through the holes. Tape to hold in place. Cover one of the big ends of the binoculars with card, and secure it with tape.

The dark spots on the image are sunspots. Draw over the spots daily and keep a record of their movements.

2 Position the binoculars so that sunlight shines through them, as shown. Move the white paper until you see an image of the Sun on it.

Sudden bursts of hot bright gas are called flares.

Eclipse

When the Moon is directly between the Sun and the Earth, it stops sunlight from reaching a part of Earth. Just as the Sun disappears behind the Moon, the eclipsed (covered) Sun looks like a giant diamond ring.

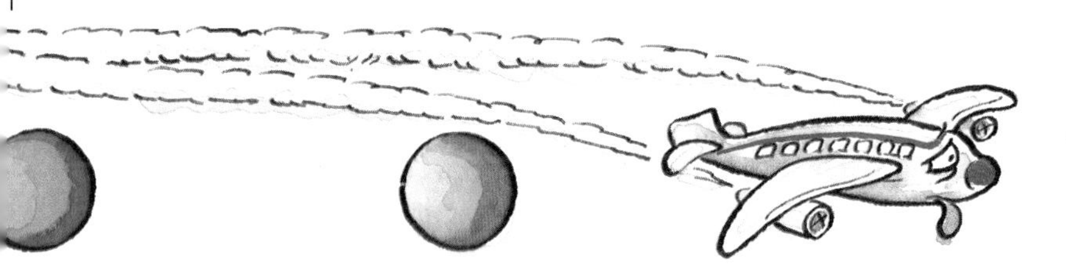

The Solar System

The Sun and its family is called the Solar System. As well as the planets, there are over 60 moons, millions of small rocks called asteroids, and billions of comets in the family. The Sun is in the centre of the Solar System – everything else travels around it.

Going in circles
Each planet follows its own path, called an orbit, around the Sun. All the planets travel in the same direction, but at different speeds from each other.

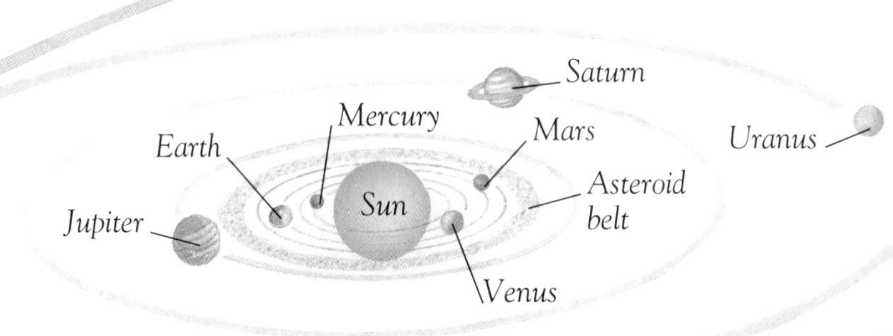

In a spin
A planet's year is the time it takes to orbit the Sun once. Pluto's year is the longest – it lasts for 248 Earth years. The planets spin as they orbit the Sun. A planet's day is the time it takes to spin around once.

Big and small

There is a big difference in size between the planets. Jupiter is about 63 times bigger than Pluto. Pluto is the smallest planet. Next smallest is Mercury, followed by Mars, Venus, Earth, Neptune, Uranus, Saturn, and the biggest – Jupiter.

Planets on your walls

Bring the Solar System to your bedroom. All you need is thin card or paper, colouring pencils or pens, a tape measure, and scissors.

Pluto

Outside planet

Pluto is the farthest planet from the Sun. But for part of its orbit, it gets closer to the Sun than Neptune.

Neptune

1 Cut out circles of card or paper for the Sun and the nine planets. You can use different-sized round objects, such as plates and bowls, to draw around. Draw the planets on the circles, using the pictures later on in this book as a guide.

Neptune and Pluto are so far away from the Sun they may have to go outside your bedroom!

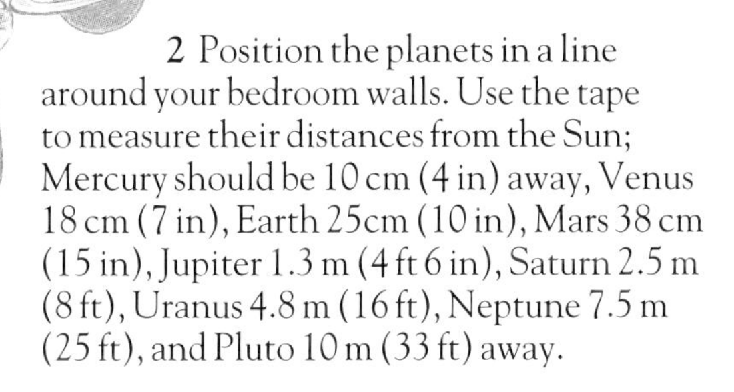

2 Position the planets in a line around your bedroom walls. Use the tape to measure their distances from the Sun; Mercury should be 10 cm (4 in) away, Venus 18 cm (7 in), Earth 25cm (10 in), Mars 38 cm (15 in), Jupiter 1.3 m (4 ft 6 in), Saturn 2.5 m (8 ft), Uranus 4.8 m (16 ft), Neptune 7.5 m (25 ft), and Pluto 10 m (33 ft) away.

Mercury and Venus

The two closest planets to the Sun, Mercury and Venus, are both rocky planets, but visitors would find them very different from each other. Mercury is an almost airless world with no atmosphere (layer of gas around it). Venus is surrounded by a very thick layer of clouds.

Hot and cold

Mercury is scorching hot by day. But without an atmosphere it cannot hold in its daytime heat, so at night it is unbearably cold.

Bright planet

Venus shines brightly because the clouds that surround it reflect sunlight well. Venus often looks like a very bright star in Earth's sky just before sunrise and just after sunset. Because of this, it is known as the morning or evening star.

Venus's clouds stop us from seeing its rocky surface.

Not for humans

Venus is not a friendly place for humans to visit. If you stepped on to Venus's surface, its hot temperature would bake you, the gas in the clouds would poison you, and the pressure of the gas would be so strong it would crush you.

Battered planet

Mercury is covered with bowl-shaped hollows called craters. Mercury's craters were formed millions of years ago when asteroids crashed into it. It is not easy to spot Mercury in the sky because it stays close to the Sun. You may sometimes see it near the horizon (where the Earth meets the sky) at sunrise or sunset.

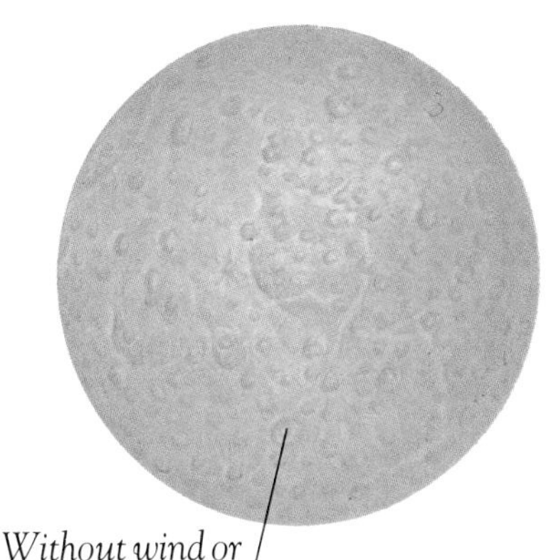

Without wind or water, Mercury's craters will never wear away.

Making craters

Try making your own crater landscape. You will need a washing-up bowl, flour, and objects of different shapes and sizes, such as a marble, a stone, or a ball of modelling clay. These objects will be your "asteroids".

1 Pour some flour into a bowl, spread it out evenly, and smooth down the surface with a spoon. The flour should be about 5 cm (2 in) deep. This will be your "landscape".

2 Place the bowl on some newspaper on the floor. Stand on a stool or chair and drop your asteroids onto your landscape.

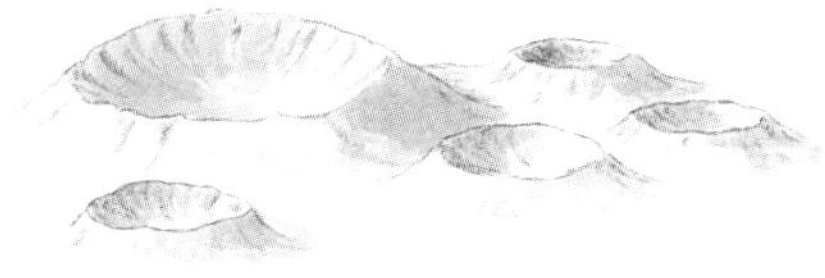

Many of the planets and moons have craters.

3 See how different-sized asteroids make different-sized craters. Bigger asteroids make wider and deeper craters. Try throwing the asteroids at different speeds. The faster they travel, the more impact they have on the landscape.

Wear an apron to keep your clothes clean, and use lots of newspaper to avoid making a mess on your floor!

Planet Earth

There is no other planet like Earth. As far as we know, it is the only planet where there is life. Earth hasn't always looked like it does today, nor will it stay the same. Volcanoes, earthquakes, the weather, and humans, all change the Earth in different ways.

A heavy load

In Greek mythology, a giant called Atlas rebelled against the gods. The gods punished him by making him hold up the Earth's sky forever.

Journey to Earth

Imagine travelling to Earth from far away in space. At first, it would look like a blue planet. But as you travelled closer, you would see white clouds, and then brown land and oceans of water.

Earthshine

If you were on the Moon, the Earth would shine and move across the sky just like the Moon does in our sky. The part of Earth you see here is in daylight. It is night for the other, dark side.

Water covers three-quarters of Earth's surface. Earth is the only planet with water.

The Earth is like a spaceship travelling through space. It is moving at about 107,000 km/h (66,500 mph) as it orbits the Sun. That's more than 100 times faster than a jumbo jet!

Bright pink and blue light glows across the sky.

Light fantastic
The Sun blasts out particles (tiny pieces) which enter Earth's atmosphere and create amazing light displays. These displays can be seen in the sky towards the North and South Poles. They are known as the Northern and the Southern Lights.

There were no plants and animals for millions of years. Dinosaurs lived on Earth 200 million years ago.

Modern Earth is dominated by humans and their machines.

The Earth is made mostly of rock and metal.

The land on Earth used to be in one large lump. It gradually split and the pieces moved apart.

Story of the Earth
Earth is about 4,600 million years old. In the beginning, the Earth was cold. It warmed up slowly and became so hot that most of its metals and rocks melted. The metals sank to the Earth's centre and the rocks floated on top. As the Earth cooled, steam fell as rain and formed the oceans.

The changing Moon

The Moon orbits the Earth as the Earth orbits the Sun. We can see the Moon most nights, and often during the day, too. But the Moon we see seems to change shape in a $29\frac{1}{2}$-day cycle. The different shapes of the Moon we see are called its phases.

Moon views
As the Moon orbits the Earth, our view of it changes as the Sun lights up more or less of the side we see. At times, it seems to disappear altogether.

Last Quarter – it is on the last quarter of its orbit.

Crescent – it will soon disappear again.

New Moon – we cannot see it as none of the side which faces us is lit up.

The Sun lights up the Moon and the Earth.

Crescent – the Sun begins to light up our side of the Moon.

First Quarter – it has moved around the first quarter of its orbit.

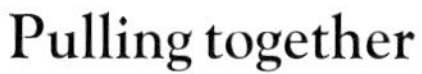

Pulling together
The pull of the Earth's gravity keeps the Moon in its orbit. The Moon's gravity pulls on the Earth's oceans, causing their high and low tides.

Phase yourself

You can create the phases of the Moon yourself. You will need a torch, silver foil, and a large and a small round object – an apple and a golf ball would do. Cover the golf ball with silver foil – this is your "Moon". The apple is your "Earth" and the torch is your "Sun".

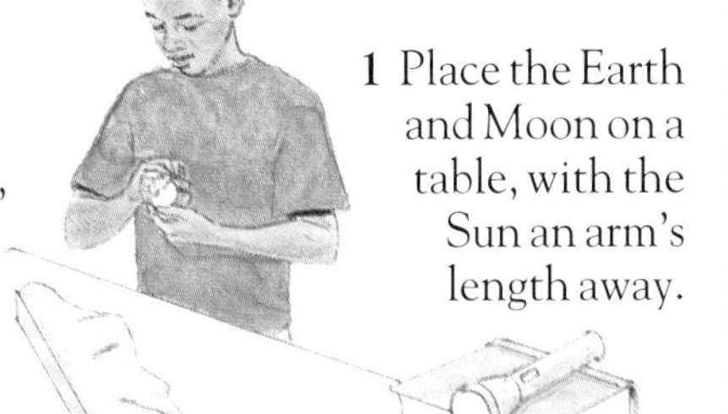

1 Place the Earth and Moon on a table, with the Sun an arm's length away.

Do this project in the dark for a better effect.

2 Move the Moon around the Earth. Stay in the same place as you do this. You'll see that different torchlit parts of the Moon – its phases – come in and out of view.

Gibbous – when the Moon gets "smaller" it is said to be waning.

Full Moon – all of the side facing us is lit up.

Gibbous – three-quarters is visible. The "growing" Moon is said to be waxing.

Moon moves

Everything in the sky moves. The Moon follows a path across the sky, and if you look at it from time to time one evening, you'll see how its position changes. In this view (right), less than one hour has passed between the top and bottom Moons.

The Moon in close-up

You can see surface details on the Moon – even in the daytime – and you do not need any special equipment. Because the Moon has no atmosphere, there are no Moon clouds to spoil your view. The dark patches you see are lowlands; the brighter areas are highlands.

The Moon's craters, like Mercury's, were made by space rocks crashing into it thousands of millions of years ago.

Moon watch

As long as the weather is fine, you will be able to see the Moon regularly and record its changing phases. You can also make drawings of its landscape. Once you have studied the Moon with your eyes, look at it through binoculars to see even more detail.

There is no air to breathe on the Moon, so astronauts have to take air with them.

We always see the same side of the Moon – this side.

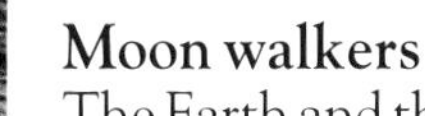

Moon walkers

The Earth and the Moon are the only places on which humans have walked. Twelve men have been to the Moon and returned home safely. The next time astronauts visit the Moon, they will probably set up a base camp where visitors can stay.

High jumps

Everyday life on the Moon would be very different from life on Earth. There would be no noise, because sound cannot travel without air. Also, the Moon's gravity is much weaker, so you could jump almost six times higher than on Earth.

A Moon of your own

Make a model of the Moon. You will need newspaper, one cup of flour, a ball, plastic film (to cover the ball), string, and paint.

Cover the ball with plastic film first.

1 Mix the flour with three cups of water. Tear the newspaper into small pieces and soak a few pieces at a time in the mixture. Cover half the ball with the wet paper. When completely dry, remove the paper half-Moon from the ball. Make a second half-Moon.

Cover with four layers of paper.

2 Attach one end of the string to the inside of one half-Moon. Put the two half-Moons together and secure with sticky tape. Cover the join with three layers of wet newspaper.

Hold the two halves tightly together.

You can make craters and highlands with more wet paper.

3 When dry, use your own Moon sketches or the pictures in this book to help you paint your Moon. Use the string to hang it up.

Mars

If you were going to choose a planet to visit, Mars would be the best place to go. It is the next planet from the Sun after the Earth, so it is not too far away. Its surface is the most like the Earth, although it is very cold on Mars. There is no air to keep animals and plants alive; red rocks and dust are everywhere; and even the sky is red.

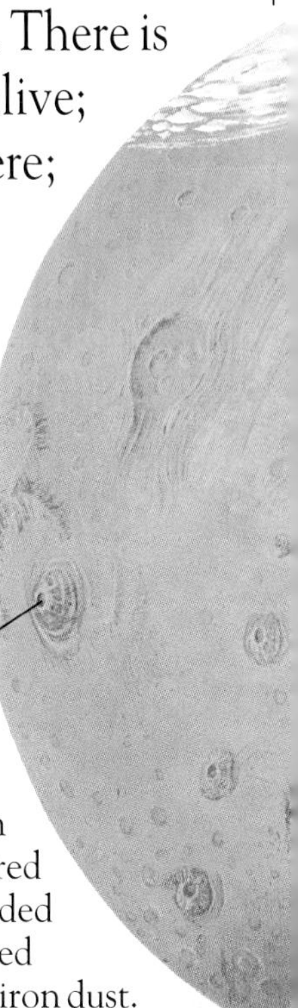

Life on Mars

As recently as one hundred years ago, some scientists believed that intelligent life lived on Mars. But we now know that nothing lives on this planet.

Olympus Mons

Violent volcano

Mars has giant volcanoes on its surface. Long ago, they erupted and helped change the surface of Mars – but today they are all dead. The biggest volcano is Olympus Mons. It is three times bigger than the highest mountain on Earth.

Dust and rust

Why is Mars called the red planet? Because from the Earth, it looks like a red disc! Spacecraft that landed on Mars found that the red colour comes from rusty iron dust.

Olympus Mons

A day on Mars lasts for nearly 25 hours, a little longer than Earth's day. But its year lasts 687 days.

The red dust covers everything and forms dunes, just like sand dunes on Earth.

The Martian air is always freezing cold.

Viking photo

This photograph of Mars's barren, rocky surface was taken by the American spacecraft *Viking 2*. Scientists are now planning to send an astronaut to Mars. He or she should be on Mars when you are an adult. Perhaps you will become an astronaut, and it could be you!

Storms of dust

Mars has dust storms which can last for weeks. Strong winds pick up the red dust and move it about the surface. With powerful telescopes, we can see the change in the colour of Mars as the dust moves around.

Ice cap

Mars has ice caps on its north and south poles.

Swirling red dust

Jupiter

Gigantic Jupiter is made mostly of gas and liquid, with a small rocky core (centre). Jupiter has a cloudy outer layer with dark bands and bright zones of different colours. The cloud tops reflect sunlight well, so the planet shines brightly in our sky.

Jupiter's journey

Jupiter can be mistaken for a bright silver-white star in the night sky. It moves slowly against the starry background, and you can watch its progress from week to week. It takes about a year to travel through each zodiac constellation.

Europa – one of Jupiter's moons

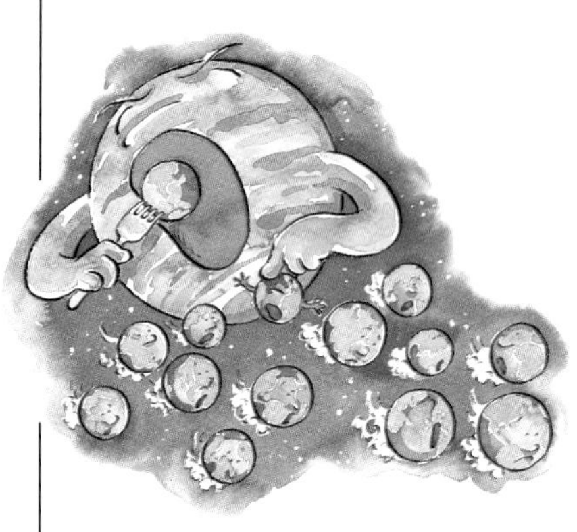

Earth meal

The Earth is tiny compared to Jupiter. It would take 1,330 Earth-sized mouthfuls to fill up Jupiter.

Stormy weather

Jupiter spins around very fast. This produces high-speed winds and terrific storms. These are much bigger and longer-lasting than storms on Earth. The Great Red Spot storm has been raging for at least three hundred years!

The Great Red Spot is the biggest storm in the Solar System.

Jupiter's cloud layer is icy cold. But the planet gets hotter and hotter towards the centre.

Sea of liquid hydrogen

Metal hydrogen

Rocky metal core

The cloud layer is 1,280 km (800 miles) thick.

Cross-section
Jupiter's layer of clouds is made mostly of hydrogen gas. Underneath the layer of clouds is a sea of liquid hydrogen. Beneath the sea is more hydrogen, but this time it is in the form of metal. At the centre is the planet's rocky metal core.

Many moons
Jupiter has sixteen moons. The four biggest – Ganymede, Callisto, Io, and Europa – can be seen from Earth with binoculars.

Callisto

Ganymede

Jupiter has a faint ring that was discovered by spacecraft in 1979. The ring cannot be seen from Earth, even through the most powerful telescopes.

Io

King god
We don't know who named the planets, but Jupiter is a fitting name for this planet. Jupiter was king of the gods of Ancient Rome; and the planet Jupiter, being the biggest, is the king of all the planets!

If you would like to visit Jupiter on your next school trip – you're out of luck. Jupiter is so far away it would take a bus around 1,000 years to get there!

Saturn

Like Jupiter, Saturn is a giant planet made mainly of gas and liquid. It is surrounded by a system of rings which stretch for thousands of kilometres into space. Saturn also has the largest family of moons in the Solar System.

Floater
Saturn's material (the stuff it is made of) is not as tightly packed as the other planets. If Saturn could be placed in water, it would float!

Colourful clouds
Saturn's clouds are coloured by the different materials in them. Scientists have looked through holes in the top layer of gold-coloured clouds, and seen brown and blue clouds below.

The rings can be seen only through a telescope.

Big ears!
When astronomers first saw Saturn's rings nearly four hundred years ago, they didn't know what they were. Saturn looked as though it had ears!

Saturn's rings are not solid. They are made of millions of pieces of rock and ice.

Winds ten times faster than hurricanes on Earth whizz around Saturn.

View of the rings

Our view of Saturn's rings changes as it orbits the Sun. Every 15 years Saturn is sideways to us and the rings seem to disappear.

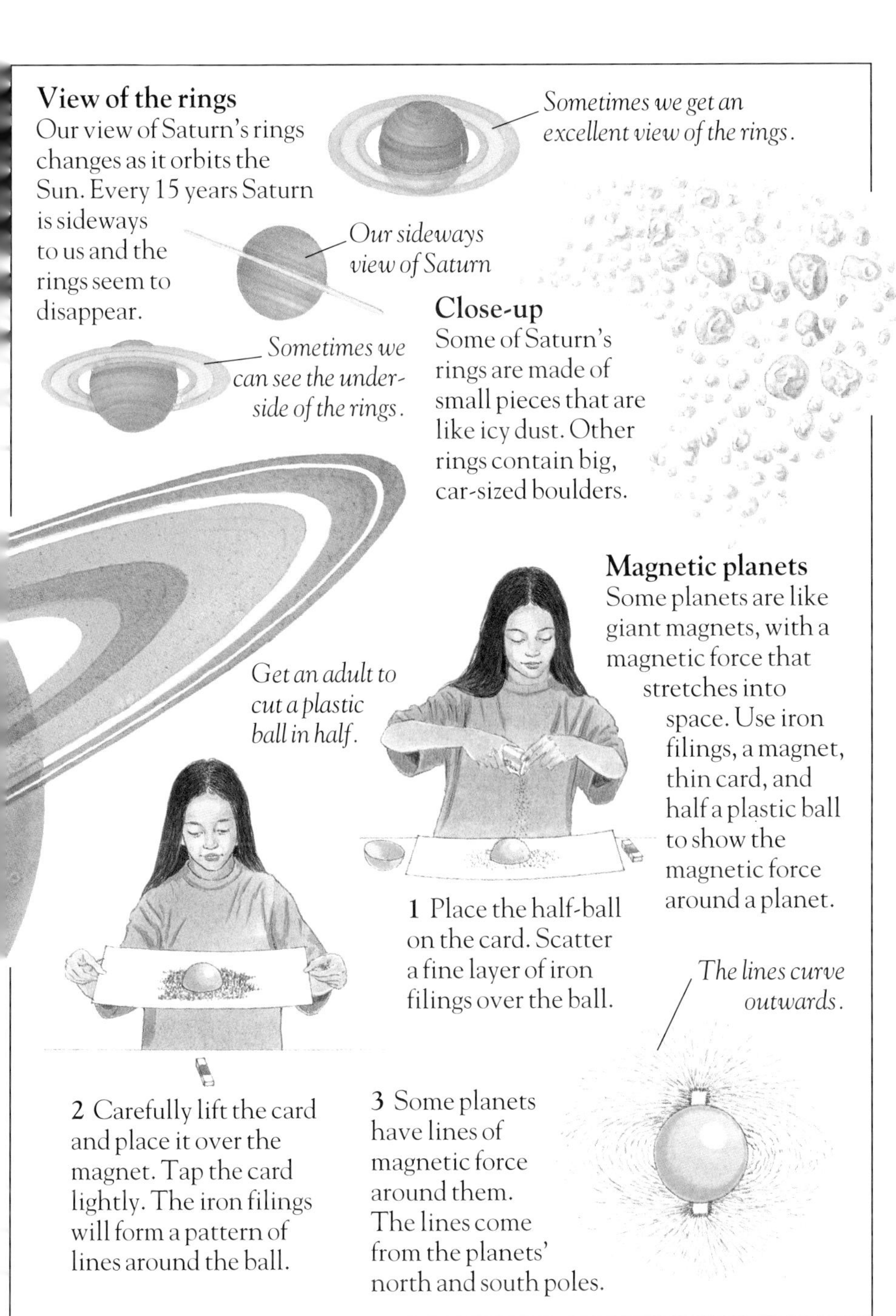

Sometimes we get an excellent view of the rings.

Our sideways view of Saturn

Sometimes we can see the under-side of the rings.

Close-up

Some of Saturn's rings are made of small pieces that are like icy dust. Other rings contain big, car-sized boulders.

Magnetic planets

Some planets are like giant magnets, with a magnetic force that stretches into space. Use iron filings, a magnet, thin card, and half a plastic ball to show the magnetic force around a planet.

Get an adult to cut a plastic ball in half.

1 Place the half-ball on the card. Scatter a fine layer of iron filings over the ball.

2 Carefully lift the card and place it over the magnet. Tap the card lightly. The iron filings will form a pattern of lines around the ball.

3 Some planets have lines of magnetic force around them. The lines come from the planets' north and south poles.

The lines curve outwards.

Uranus and Neptune

These two giant planets are so far away that you need powerful telescopes to see them. We have learned most of what we know about Uranus and Neptune from the space probe *Voyager 2*. Both planets are made mostly of gas and liquids, and are very, very cold.

Uranus's "ring" is made up of 11 separate rings of dark rock.

Side spin

Uranus's outer layers are made mostly of hydrogen and helium gas. Another gas, called methane, gives the planet its blue-green colour. Uranus spins on its side as it orbits the Sun. A system of rings and 15 moons orbit around Uranus's middle.

Uranus's summer lasts for 42 years! But summer on Uranus is very cold.

Voyager 2 *sent thousands of pictures back to Earth.*

A long journey

Voyager 2 left Earth in 1977 to travel to the distant giants of the Solar System. After visiting Jupiter and Saturn, it reached Uranus in 1986 and Neptune in 1989.

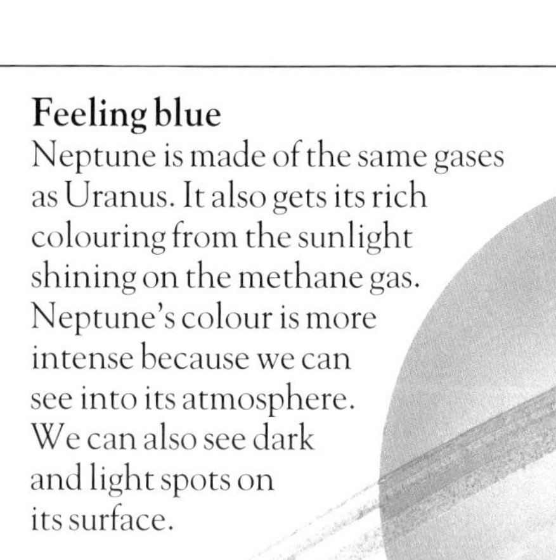

Feeling blue

Neptune is made of the same gases as Uranus. It also gets its rich colouring from the sunlight shining on the methane gas. Neptune's colour is more intense because we can see into its atmosphere. We can also see dark and light spots on its surface.

Neptune's rings are very dim.

Neptune is the windiest planet in the Solar System.

The Great Dark Spot is a huge storm which is almost as big as planet Earth.

White spots are clouds of methane ice.

Racing round

Neptune's dark and white spots all travel around the planet at different speeds. Sometimes they are close together; at other times they are apart. One white patch travels so fast that it has been nicknamed "the scooter".

Order of the planets

Make up a sentence using the first letter of each of the planets' names to help you remember their order out from the Sun. Start with "M" for the first planet – Mercury: for example, **M**ost **V**iolet **E**vening **M**oths **J**itter **S**lowly **U**ntil **N**ight **P**asses.

Pluto, moons, and asteroids

The planet Pluto is so small and so distant from Earth that we cannot get a good look at it – even with the most powerful telescopes. Pluto has one moon called Charon, which is about half Pluto's size. There are 61 other moons in the Solar System, 18 of them living with Saturn. Mercury and Venus are the only planets without any moons.

Small world
Pluto is more like a moon than a planet. It is really tiny – much smaller than Earth's Moon. Pluto is a freezing and dark world of ice and rock. The Sun is so far away that Pluto gets hardly any heat and light from it.

Some scientists think Pluto was once a moon of Neptune that got knocked out of its orbit.

Charon takes just over six days to travel once around Pluto.

Charon
Pluto was discovered in 1930, but scientists did not discover its moon until 1978. Charon is probably rocky and icy, like Pluto. Spacecraft haven't visited Pluto and Charon yet. Until they do, Pluto and its moon remain worlds of mystery.

From Pluto, the Sun would look like just another bright star.

Scientists can tell what the surface of a planet or moon is like by studying sunlight reflecting on its surface.

Many moons

Moons do not all look the same. Ganymede, one of Jupiter's moons, is icy and covered in craters; Europa is smooth and covered with dark lines; and eruptive Io looks like a giant pizza!

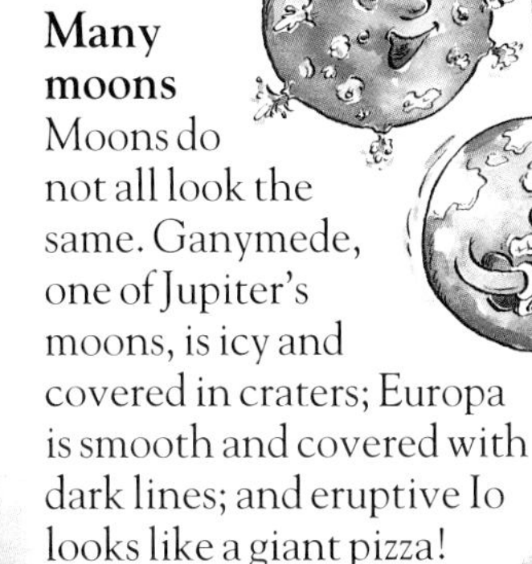

Reflecting light

Planets and moons have no light of their own; they are lit by the Sun. Using a torch, shine light on different objects, such as an egg, a ball covered in silver foil, and some coal. See how some objects reflect light better than others. Try this in a darkened room to make it more like space.

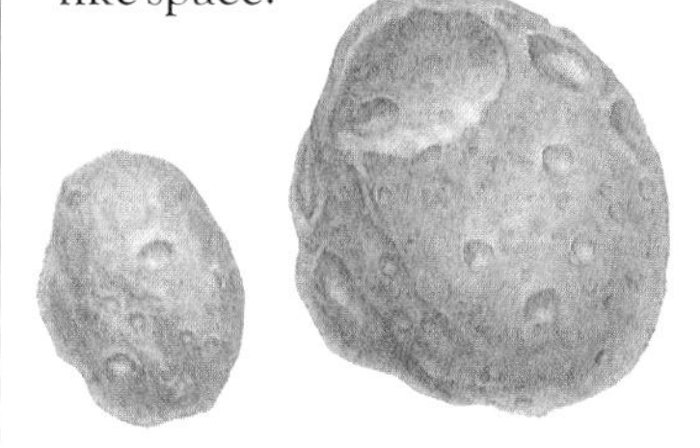

Moons or asteroids?

Mars has two tiny moons, Phobos and Deimos. They are potato-shaped, just like some asteroids. It is possible they were once asteroids which were captured by Mars.

Asteroid belt

There are millions of small rocky bodies called asteroids in the Solar System. Most live in between Mars and Jupiter and travel around the Sun together. They form an enormous belt around the Sun.

Comets and meteors

Comets are enormous balls of snow and dust, like gigantic dirty snowballs. They live together in a huge cloud at the edge of our Solar System. Occasionally, a comet is knocked out of the cloud and starts a journey towards the Sun. You may see one in the night sky as it passes by Earth – it will look like a fuzzy star.

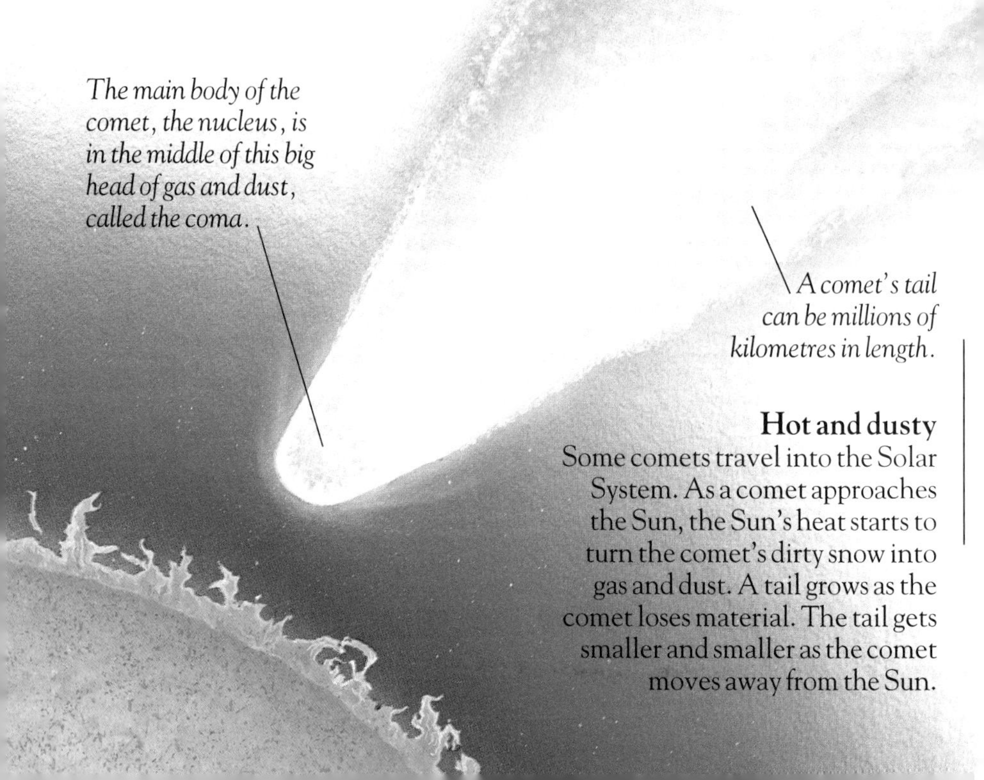

The main body of the comet, the nucleus, is in the middle of this big head of gas and dust, called the coma.

A comet's tail can be millions of kilometres in length.

Hot and dusty

Some comets travel into the Solar System. As a comet approaches the Sun, the Sun's heat starts to turn the comet's dirty snow into gas and dust. A tail grows as the comet loses material. The tail gets smaller and smaller as the comet moves away from the Sun.

Halley's return

Some comets return again and again to Earth's sky. A comet called Halley's Comet orbits the Sun once every 76 years or so. Comets like Halley's don't last forever. They lose material as they travel. When Earth moves through this material, we see a meteor shower in the night sky.

Comet cure

Earth passed through the tail of Halley's Comet in 1910. In North America, a businessman made money by selling "comet pills" to people who thought the gas in the tail would poison them!

A light shower

When you see a meteor shower, streaks of light cross the night sky as the comet material burns up in our atmosphere. Meteors are also called shooting stars. Look out for the Geminids shower which occurs around 7-15 December every year. You can see up to 50 shooting stars in an hour.

Arizona crater

Car-sized space rocks are too big to burn up in the atmosphere; they crash into Earth's surface. These are called meteorites. The big Barringer meteorite made this 1.3-km- (0.8-mile-) wide crater in Arizona in the United States around 25,000 years ago.

Astronomers' tools

Astronomers use special tools to study the stars and planets. The telescope is the most important tool for finding out about the night sky. But astronomers do not look through telescopes very often. Modern telescopes work on their own, recording what they see. This information is then passed on to the astronomer, who works in an office.

The observatory's dome-shaped roof slides open when the telescope is to be used.

Milky Way

Telescopes like this one can take photographs.

The Large Magellanic Cloud

The Milky Way shows up well in the sky above this observatory in Chile, South America.

Above the clouds
Telescopes are kept in dome-shaped buildings called observatories. The world's best observatories are placed high on mountain-tops. This places the telescopes above the clouds and keeps them away from city lights, which would spoil the view.

The Magellan space probe orbits Venus. It sends radio messages to Earth which show what Venus's surface is like.

A great help

Computers control telescopes, satellites, and space probes, and also help astronomers work out what they have seen. In modern astronomy, computers are just as important as telescopes.

Space robots

Earth's atmosphere stops some information from space reaching us. To overcome this problem, astronomers send their tools into space. Robots, called space probes, are sent to get a closer look at the planets. Other robots, called satellites, orbit Earth above the atmosphere and look out into space.

The Hubble Space Telescope satellite orbits Earth. It looks out into space, seeing farther than any of the Earth-based telescopes.

There are observatories on mountain-tops all around the world.

Special telescopes, called radio telescopes, listen in to the stars.

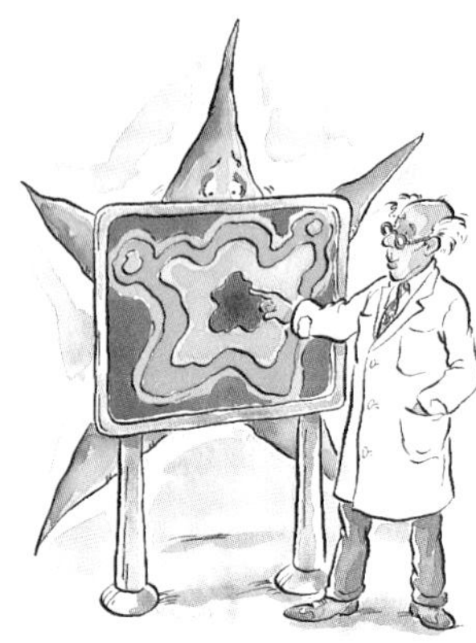

Seeing into a star

Different tools look at the stars in different ways. X-ray satellites take X-ray pictures. A star may not shine very brightly, but an X-ray picture might show that there is a lot of activity in the star, and that it has lots of energy.

Exploring space

We can learn a lot about what is in space by watching from Earth. But by travelling into space, we can get a better look and even see things that are invisible from Earth. Astronomers regularly send out satellites and space probes. Astronauts – human space travellers – have also explored space close to Earth.

Galileo

A space probe called *Galileo* was sent on a mission to Jupiter in 1989. It will take *Galileo* six years to reach Jupiter, even though space probes travel very fast. Once at Jupiter, *Galileo* will spend two years touring around the giant planet and its moons.

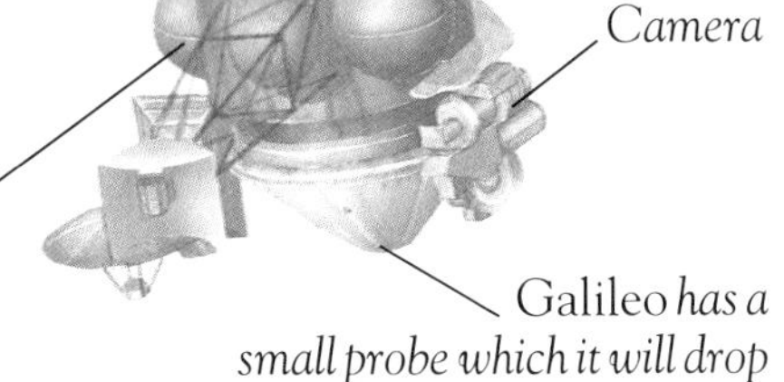

The "umbrella" part is used to communicate with Earth.

Camera

The main part of Galileo *is about the size of a bus.*

Galileo *has a small probe which it will drop into Jupiter's atmosphere.*

Getting the message

Astronomers use their computers to "talk" to the computers on space probes. Space probe computers send back what the probe has found out. Back on Earth, big dishes work like enormous ears, collecting the radio messages from the probes.

Jupiter's magnetic force will be measured by this long arm.

A pack on the back

When astronauts are outside their spacecraft they must be careful not to fly off into space. Sometimes they wear backpacks which contain small rockets, allowing the astronauts to control their movements.

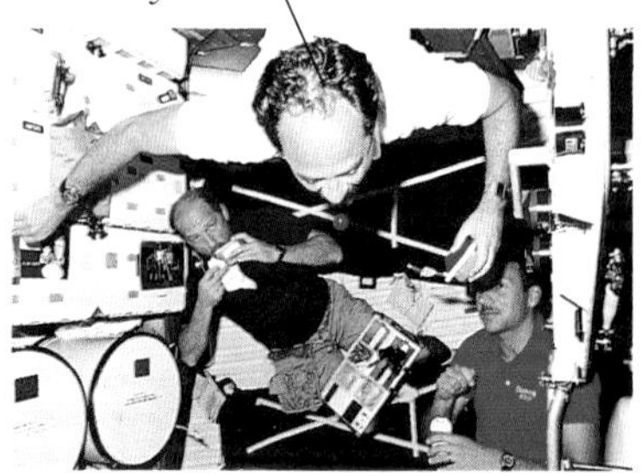

Astronaut chasing a ball of strawberry drink

Catch that drink

Everything is almost weightless in space. Things have to be tied down in case they float away. Drinks are sucked through straws, and if any escapes, the drink forms into small balls which float like little balloons.

Lift off

Rockets and space planes, called space shuttles, carry satellites, space probes, and astronauts into space. Use a balloon to see how a rocket or a space shuttle blasts off from Earth.

1 Blow up the balloon but don't tie the end – hold it between your fingers instead.

Air rushing backwards pushes the balloon forwards.

2 Let the balloon go; the air rushing out of the balloon makes it move forwards. When a rocket or a shuttle is launched, a stream of hot gases spurts out of its "tail" down towards Earth. The rocket is pushed upwards and blasts off into space, just like the balloon "blasting" away from you.

Index

The Milky Way

The planet Jupiter

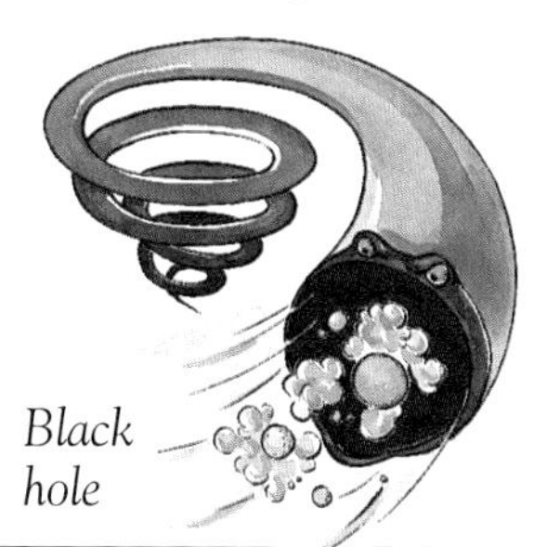

Black hole

The Scorpion

Stars have gravity and pressure

Finding Polaris

Acknowledgments

Dorling Kindersley would like to thank:
Lynn Bresler for the index.
David Hughes for checking the text.

Illustrations by:
Daniel Pyne,
Julie Anderson,
Nick Hewetson.

Picture credits
t=top b=bottom c=centre l=left r=right
Galaxy Picture Library: 41br.
Tim Ridley: 8tr.
Science Photo Library: NASA 27bl, 38bl, 42bl, 45tr, 59cl, 59tr;/ NOAO 13br (two), 29tr;/Fred Espenak 33b;/Pekka Parviainen 4bl, 39tr;/Roger Ressmeyer 56;/ Philippe Plailly 57t;/Royal Observatory, Edinburgh/ AATB 26br, 29tl;/John Sandford 55bl;/Dr. Rudolf Schild 27tr.
Anthony Thomas: 36cl.